CARNIVORES

in the Food Chain

ALICE B. McGINTY

Photography by DWIGHT KUHN

The Rosen Publishing Group's

PowerKids Press™

New York

To my mother, Linda K. Blumenthal—Alice McGinty
To Ken—Dwight Kuhn

Published in 2002 by The Rosen Publishing Group, Inc.
29 East 21st Street, New York, NY 10010

First Edition

Book Design: Maria Melendez
Project Editor: Emily Raabe
Photographs © Dwight Kuhn; p. 15 photograph © David Kuhn

McGinty, Alice B.
Carnivores in the food chain / Alice B. McGinty.
    p. cm. — (The Library of food chains and food webs)
Includes bibliographical references (p.   ).
ISBN 0-8239-5754-3 (lib. bdg.)
1. Carnivora—Ecology—Juvenile literature. 2. Food chains (Ecology)—Juvenile literature. [1. Carnivores. 2. Food chains (Ecology)
3. Ecology.] I. Title. II. Series.
QL737.C2 M219 2002
599.7'17—dc21
                                                                        00-013029

Manufactured in the United States of America

Contents

Food Chains and Webs

Carnivores sometimes form several links in a food chain. For example, after a weasel eats a rabbit that ate a plant, an owl might eat the weasel. That would form another link in the food chain. Then, if a wolf ate the owl, yet another link would be formed.

Every living thing needs energy. Living things get energy from the food they eat. Some animals, such as rabbits, eat plants. Animals that eat plants are called herbivores. Other animals, such as weasels, eat animals. Animals that eat other animals are called carnivores. If a weasel ate a rabbit that ate plants, the plants, rabbit, and weasel would form a food chain. Each time one living thing eats another, it forms a link in a food chain.

Most animals belong to more than one food chain. For example, weasels eat mice, squirrels, and snakes, as well as rabbits. Some weasels even eat porcupines! Weasels, in turn, are eaten by foxes, hawks, and owls. This makes weasels and rabbits part of many food chains. Food chains that are linked together are called food webs.

This short-tailed weasel, or ermine, as it is also known, is eating a bird to get energy. The ermine is white in the winter to blend in with the snow. This helps it hide when it is hunting. In the summer, the weasel's fur will turn brown to blend in with dirt, dead grass, and leaves.

Energy in the Ecosystem

Zebras must eat many plants to have enough energy to survive. When a lion eats a zebra, it gets energy from the zebra. Most of the energy from the plants that the zebra has eaten has already been used by the zebra to grow and move. The lion must eat many zebras to get enough energy to survive.

Together with the air, water, and land in which they live, the plants and animals that live in an area form an **ecosystem**. An ocean, a forest, or even a puddle can be an ecosystem. There are many food chains in every ecosystem. Each food chain begins with **producers**. Plants are producers. They use the energy from sunlight to produce food. The next links in each food chain are the **consumers**. Primary consumers, such as zebras, eat plants. Then secondary consumers, such as lions, eat zebras. Lions are at the top of their food chain. That means that no other animals eat lions. When a lion dies,

decomposers break up its body. Decomposers are the last link in every food chain.

The lion has a big, muscular body to help it hunt and kill other animals, such as giraffes or zebras. It has powerful legs for chasing, strong jaws for biting, and an alert brain to help it outsmart the animals that it hunts and eats.

Predators and Prey

Animals that hunt and kill other animals are called **predators**. The animals that predators hunt are called **prey**. Carnivores are predators. They must hunt and kill the food they eat.

In the insect world, the praying mantis is a predator. To survive, the praying mantis must be a good hunter. It has powerful front legs, which it uses to grab other insects, such as grasshoppers. The praying mantis holds out its front legs as though it is praying. It is actually waiting for an insect to come along. When one does, the mantis grabs and eats it. Praying mantises have good eyesight so they can see their prey and grab it quickly. They

Sometimes praying mantises eat other praying mantises. Animals that kill and eat members of their own kind are called cannibals. Cannibalism helps the strongest members of the species survive. Other animals that are cannibals are spiders and horned frogs.

also have strong mouthparts to bite their prey.

Praying mantises are also prey. They are food for other carnivores, such as birds and large spiders. The praying mantis uses **camouflage** to hide from predators that might hunt it. The praying mantis' body is thin and brown or green like the twigs on which it lives.

Ladybugs eat tiny, plant-eating insects called aphids. Ladybugs are important predators. Because they eat many aphids, they ensure that the aphids will not eat too many plants.

How Carnivores Hunt

Some prey animals protect themselves with their size. Moose, giraffes, and elephants, for example, are big and strong so they can defend themselves against predators who would like to eat them.

Each kind of carnivore has found successful ways to hunt its prey. Lions and wolves hunt in groups and surround their prey. Cheetahs run quickly, up to 60 miles per hour (97 km/h), to chase prey. The deer or antelope that cheetahs chase can run only 40 miles per hour (64 km/h). The polar bear uses its strong body to overpower the seals and caribou that it hunts. Tigers and leopards use camouflage to hide in the grass or in trees. They sneak up and pounce on wild pigs and deer. Dogs and wolves use their sensitive noses to smell prey. The wildcat has excellent hearing and eyesight to help it find mice, birds, and hares. Spiders sense movement in the air or on the ground when their insect prey is near. Prey protect themselves from predators as well. Zebras graze together in herds. Snakes, birds, deer, and other animals use camouflage to hide from predators. Skunks and weasels give off a bad smell to make predators go away.

The crab spider uses camouflage to capture its prey. When an insect lands on the flower, the crab spider pounces on it and bites it with its fangs.

This garden spider has trapped a grasshopper in its sticky web.

Carnivores Come in All Sizes

All living things are made of **cells**. Large plants and animals have millions of cells. **Organisms** with small numbers of cells are tiny. They are called **microscopic** organisms.

Zooplankton are microscopic animals that live in oceans and lakes. Some zooplankton are carnivores. Zooplankton that are carnivores eat other zooplankton. These tiny carnivores capture their prey in many ways. Some keep their mouths wide open to catch zooplankton floating by. Others have **tentacles**, which they use to sweep prey into their mouths. Still other zooplankton have filters to strain their prey from the water. Tiny carnivores are

An ocean food chain may work as follows: Some tiny water plants are eaten by a water flea in the zooplankton. The water flea is eaten by a seahorse. The seahorse is eaten by a big fish. The big fish is eaten by a shark. The shark is at the top of this ocean food chain. However, sometimes sharks are caught and eaten by people.

often food for larger carnivores. Food chains in the ocean can have many links.

This tiny predator is called a hydra. It is capturing and eating a water flea. Hydras live in fresh water, such as lakes, ponds, or streams. Their tube-shaped bodies are about 1 inch (2.5 cm) long. They have long tentacles with stingers, which they use to sting their prey.

Carnivores in the Sea

The moray eel uses its sense of smell to locate prey. It has strong teeth in the back of its mouth to crush the hard shells of clams and lobsters. In the front of its mouth, the moray eel has fanglike teeth that angle backward so the fish it swallows cannot escape. Most moray eels grow to be 4 or 5 feet (1.2–1.5 m) long.

Carnivores that live in the sea hunt in many ways. Clams filter their food from the seawater. Clams have hard shells to protect themselves from predators. Starfish, however, use suction cups on their arms to pull open the clamshells. Then the starfish sticks its stomach into the clamshell to have its meal. At the dark bottom of the sea, the deep-sea anglerfish dangles a small light at the end of its nose. When curious fish swim close to the light, the anglerfish eats them.

The largest carnivores, such as killer whales and sharks, are at the top of most ocean food chains. They eat seals, squid, fish, and many other sea creatures. Killer whales, which are also called orcas, hunt in groups and circle their prey.

Every animal in every ecosystem must do two things to survive. First, it must find food. Second, it must protect itself from being eaten by other predators. This lionfish has long, poisonous spines to protect itself from predators. Its bright colors warn predators to stay away.

Reptile Carnivores

Reptiles are animals that move on their bellies, either by sliding or crawling. Many reptiles are carnivores. Snakes, crocodiles, turtles, and lizards are reptile carnivores. One kind of lizard, called the chameleon, catches crickets and other insects with a quick flick of its long tongue. Chameleons' bodies sometimes change color depending on light or temperature. They might also change color if they are frightened or angry. Other insect hunters, such as frogs and toads, also have sticky tongues that they use to catch insects.

Some snakes, such as cobras, bite their prey with poisonous fangs. Other snakes, such as boa constrictors, **suffocate** their prey by wrapping around them so tightly

The chameleon's tongue can extend up to 5 inches (12.7 cm) in $\frac{1}{16}$ of a second! The chameleon also has good eyesight. Its two eyes can move separately in all directions to spot prey.

16

that the prey can't breathe. Snakes open their mouths wide and swallow their prey whole, without chewing. The snake's sharp teeth point toward the back of its mouth so prey cannot escape after it is swallowed.

This chameleon in Africa has just gotten itself a tasty insect meal.

Carnivores That Fly

The puffin can carry a lot of small fish in its brightly colored beak. Puffins can fly in the air and swim underwater. They use their short wings as flippers to chase fish in the cold arctic waters. Puffins live in underground homes, called burrows.

Carnivorous birds use their beaks as tools to help them hunt prey. Birds' beaks are shaped differently depending on the food they eat and the way they hunt. The kingfisher perches on a branch over water. When it sees a fish, the kingfisher dives into the water. The kingfisher has a long, spearlike beak, which it uses to stab the fish. Other fish-catching birds have hooked or sawlike beaks to help them grab slippery fish.

Birds that eat insects usually have straight, pointed beaks. The blackbird uses its beak to dig into the ground and grab insects and grubs. The woodpecker uses its beak as a hammer to make holes in trees and find insects.

Owls, hawks, and eagles are called birds of prey. They eat rabbits, mice, and many other small animals. Birds of prey have strong, hooked beaks to pull off bits of meat from their prey.

Owls have special feathers that help them fly quietly. They swoop down silently and catch their prey with curved claws. Owls carry prey back to their nests in their strong beaks. This saw-whet owl has captured a short-tailed shrew for its meal.

Mammal Carnivores

The tiger is the largest cat in the world. This carnivore can be up to 12 feet (3.6 m) long and can weigh as much as .33 tons (.29 t)! Tigers sneak up quietly on their prey. They hold a prey animal with their two front paws and kill it by biting it in the neck.

Mammals are animals that have hair and that feed milk to their young. Some mammals, such as mountain lions and weasels, are carnivores. They have long, pointy front teeth to help them grab and bite their prey. Their back teeth, called molars, are sharp also. The sharp molars help mammal carnivores tear off pieces of meat from their prey. About one-third of all mammals eat insects. These animals are called **insectivores**. The anteater is an insectivore that uses its long snout and strong claws to dig for ants. Its sticky tongue can reach 8 inches (20 cm) into an anthill. The anteater's tongue brings up hundreds of ants at a time.

Shrews eat small animals and insects. To have enough energy to survive, each day the shrew must eat enough animals to equal half of its body weight. This means the shrew must spend much of its time hunting prey. Most carnivores only catch one out of every ten animals they try to catch.

This short-tailed weasel is brown for the summer. In the winter, the weasel's fur will turn white. The weasel's body is long and thin, the perfect shape for chasing small prey into underground burrows.

Plants As Predators

Plants cannot chase their prey, so they must trap it. The sundew lures thirsty insects onto its leaves by showing off red hairs with drops of juice at the tips. These juice drops are sticky, and when the insect lands it sticks to the leaf and is trapped.

Ecologists are scientists who study predators and prey. One group of predators that interests ecologists is plant carnivores. These plants eat insects! Plant carnivores lure insects into their traps with sweet-smelling nectar, bright colors, and water.

Ecologists study the balance of predators and prey in each ecosystem. Predators keep their ecosystems balanced by eating only as much food as they need. When people kill animals for their fur or tusks, they can ruin the balance of an ecosystem. Many kinds of animals are **endangered**, or even **extinct**, because people have killed too many of them.

Glossary

camouflage (KA-muh-flahj) The color or pattern of an animal's feathers, fur, or skin that helps it blend into its surroundings.

cells (SELZ) The many tiny units that make up living things.

consumers (kon-SOO-merz) Members of the food chain that eat other organisms.

decomposers (dee-kum-POH-zerz) Organisms that break down the bodies of dead plants and animals.

ecologists (ee-KAH-luh-jists) Scientists who study the way living things are linked with each other and with the earth.

ecosystem (EE-koh-sis-tum) The way that plants and animals live in nature and form basic units of the environment.

endangered (en-DAYN-jerd) When something is in danger of no longer existing.

extinct (ik-STINKT) To no longer exist.

insectivores (in-SEK-tih-vohrz) Animals that eat insects.

microscopic (my-kruh-SKAH-pik) Something so small that it can only be seen through a microscope.

organisms (OR-geh-nih-zehmz) Living beings made of dependent parts.

predators (PREH-duh-terz) Animals that kill other animals for food.

prey (PRAY) An animal that is hunted by another animal for food.

producers (pruh-DOO-serz) Plants and algae that use sunlight to make their own food.

suffocate (SUH-fuh-kayt) To stop the flow of air to an animal's lungs.

tentacles (TEN-tuh-kuhlz) Long thin growths usually on the head or near the mouths of animals, used to touch, hold, or move.

zooplankton (ZOH-uh-plank-ton) Tiny animals that float free in water as part of plankton.

Index

Web Sites

To learn more about carnivores and food chains, check out these Web sites:
www.aza.org/gallery
www.rainforestlive.org.uk

24